The Future of Space Travel: Advancements Needed for Safe Interplanetary Travel

1. Introduction

Overview of Human Interest in Space Exploration

Since the dawn of human civilization, the stars have captivated our imagination. Ancient cultures, from the Egyptians to the Mayans, constructed elaborate myths and calendars based on the movement of celestial bodies.

The twinkling lights of the night sky represented the unknown, the vast, uncharted territory beyond our Earth. Fast forward to the 20th century, and our fascination with space took on a new dimension with the advent of modern astronomy and rocketry. The launch of Sputnik in 1957 marked humanity's first step into space, and just over a decade later, Neil Armstrong's "giant leap for mankind" during the Apollo 11 mission became a defining moment in human history.

Today, space exploration is no longer just a domain of national pride and scientific discovery. It has become a pivotal part of our quest to push the boundaries of human knowledge, to understand our place in the universe, and to ensure the long-term survival of our species. The challenges of Earth—climate change, overpopulation, and resource depletion—have sparked renewed interest in the possibility of human settlements on other planets, particularly Mars.

Visionaries like Elon Musk and Jeff Bezos are pouring resources into the development of technology that could make interplanetary travel a reality within our lifetimes.

As a result, space travel has shifted from the realm of science fiction to a critical area of research, with profound implications for the future of humanity.

Brief History of Space Travel

The history of space travel is a tale of triumphs, setbacks, and relentless human ingenuity.

The space race between the United States and the Soviet Union in the mid-20th century accelerated technological advancements at an unprecedented pace.

The launch of Sputnik was followed by a series of manned missions, culminating in the Apollo program.

These early missions not only demonstrated the feasibility of sending humans into space but also laid the groundwork for future exploration. The development of space stations, like Skylab and the International Space Station (ISS), provided insights into living and working in space for extended periods.

In the decades following Apollo, space exploration faced a lull, with missions primarily focused on low Earth orbit activities, such as deploying satellites and conducting scientific research aboard the ISS.

However, the turn of the 21st century has seen a resurgence in space exploration efforts.

Private companies like SpaceX, Blue Origin, and others have entered the arena, bringing innovation and reducing the cost of space travel through reusable rockets and new spacecraft designs.

NASA, ESA, and other space agencies have also set their sights on Mars, planning missions that could eventually lead to human colonization of the Red Planet.

Importance of Advancing Space Travel Technology for Future Missions

As we set our sights on destinations beyond Earth's orbit, the challenges of space travel become more daunting. Mars, our most likely candidate for colonization, is an average of 225 million kilometers away from Earth. A round-trip journey to Mars would take about 18 months, including the time spent on the planet's surface. This is vastly different from the few days it took Apollo astronauts to reach the Moon.

The increased distance and duration of interplanetary travel bring about new and complex challenges that must be addressed to ensure the safety and success of future missions.

Advancing space travel technology is not just about getting to other planets; it's about surviving the journey and thriving upon arrival.

This involves developing new propulsion systems that can reduce travel time, creating life support systems that can sustain human life for years, and designing spacecraft that can protect astronauts from the harsh environment of space. Radiation exposure, microgravity, and the psychological effects of isolation are just a few of the obstacles that must be overcome.

Current State of Space Travel: Space Agencies and Private Companies

Space travel has evolved dramatically since the mid-20th century, transitioning from government-led missions to a more diverse landscape where private companies play a crucial role.

This shift is driven by the need for innovation, cost-efficiency, and the ambitious goal of establishing a human presence on other planets.

Let's delve into the key players in the current space travel industry, both governmental and private, and their contributions to the future of interplanetary exploration.

1. Governmental Space Agencies

1. NASA (National Aeronautics and Space Administration)

- Overview:

 NASA has been at the forefront of space exploration since its inception in 1958.

 With historic achievements like the Apollo Moon landings and the Mars rovers, NASA remains a global leader in space technology.

- Current Missions:

NASA's Artemis program aims to return humans to the Moon by 2025 and establish a sustainable presence there by the end of the decade.

This mission is seen as a stepping stone for future Mars exploration.

Additionally, NASA is working on the Mars Sample Return mission, which will bring samples from the Martian surface back to Earth.

- Future Goals:

 NASA's long-term goal is to send humans to Mars in the 2030s.

 The agency is developing the Space Launch System (SLS) and the Orion spacecraft to carry astronauts on deep space missions.

2. ESA (European Space Agency)

- Overview:

ESA, formed in 1975, coordinates the space activities of 22 European countries.

It has been involved in numerous scientific missions, such as the Rosetta comet mission and the Mars Express orbiter.

- Current Missions:

 ESA is collaborating with NASA on the Artemis program and the Mars Sample Return mission.

 It also operates the ExoMars program, which aims to explore the Martian surface and search for signs of life.

- Future Goals:

ESA plans to contribute to lunar and Martian exploration by developing new technologies and participating in international partnerships.

3. Roscosmos (Russian Federal Space Agency)

- Overview:

Roscosmos, established after the dissolution of the Soviet Union, continues Russia's legacy of space exploration.

It was instrumental in the development of space stations, including Mir and the ISS.

- Current Missions:

Roscosmos is involved in ISS operations and has plans to launch a new space station, the Russian Orbital Service Station (ROSS).

It also continues to launch crewed and uncrewed missions to the ISS using its Soyuz spacecraft.

- Future Goals:

Roscosmos is developing the Orel spacecraft, designed for deep space missions, and has long-term plans for lunar and Martian exploration.

4. CNSA (China National Space Administration)

- Overview:

CNSA has rapidly emerged as a major player in space exploration, with achievements including the Chang'e lunar missions and the Tianwen-1 Mars mission.

- Current Missions:

China's space program has focused on lunar exploration, with plans to establish a lunar base by the 2030s.

The Tianwen-1 mission successfully placed a rover, Zhurong, on Mars, making China the third country to land on the Red Planet.

- Future Goals:

 CNSA aims to conduct a crewed mission to the Moon and has ambitions for Mars exploration, including potential crewed missions in the future.

2. Private Space Companies

1. SpaceX

- Overview:

Founded by Elon Musk in 2002, SpaceX has revolutionized space travel with its emphasis on reusability and cost reduction.

The company's Falcon 1, Falcon 9, and Falcon Heavy rockets have become the workhorses of commercial spaceflight.

- Key Achievements:

 SpaceX's Dragon spacecraft was the first commercial vehicle to dock with the ISS, and its Crew Dragon has transported astronauts to and from the ISS.

 The company's Starship, currently under development, aims to carry humans to the Moon, Mars, and beyond.

- Future Goals:

SpaceX's ultimate goal is to make humanity multi-planetary. The company is focused on developing the Starship spacecraft, which is designed for deep space missions and is central to Elon Musk's vision of colonizing Mars.

SpaceX also plans to launch the world's largest satellite constellation, Starlink, to provide global internet coverage.

2. Blue Origin

- Overview:

Founded by Jeff Bezos in 2000, Blue Origin aims to enable a future where millions of people live and work in space.

The company has developed several reusable rockets, including New Shepard and New Glenn.

- Key Achievements:

 Blue Origin has successfully launched and landed its suborbital New Shepard rocket multiple times, making it a key player in the space tourism industry.

 The company is also working on the New Glenn rocket, designed for orbital missions.

- Future Goals:

Blue Origin's long-term vision includes building space habitats, such as the O'Neill colonies, where people could live permanently.

The company is also developing the Blue Moon lunar lander, which aims to support NASA's Artemis program and eventually enable lunar colonization.

3. Boeing

- Overview:

Boeing is a veteran aerospace company with a rich history in space exploration, including contributions to the Apollo program and the Space Shuttle.

Today, Boeing is a major contractor for NASA and plays a crucial role in space infrastructure development.

- Key Achievements:

 Boeing is developing the CST-100 Starliner, a spacecraft designed to ferry astronauts to and from the ISS.

 The company is also heavily involved in the development of the SLS, which will be the most powerful rocket ever built and is key to NASA's Artemis program.

- Future Goals:

 Boeing aims to support deep space exploration through its work on the SLS and other projects.

 The company is also exploring opportunities in space tourism and commercial spaceflight.

4. Virgin Galactic

- Overview:

 Founded by Sir Richard Branson, Virgin Galactic is focused on space tourism.

 The company's SpaceShipTwo is designed to take paying customers on suborbital spaceflights, offering them a few minutes of weightlessness and a view of Earth from space.

- Key Achievements:

 Virgin Galactic has conducted several successful test flights of SpaceShipTwo, bringing the dream of commercial space tourism closer to reality.

- Future Goals:

 Virgin Galactic plans to begin regular commercial spaceflights in the near future, with a long-term vision of expanding its services to include orbital tourism and point-to-point travel via space.

5. Rocket Lab

- Overview:

Rocket Lab, founded in New Zealand by Peter Beck, is a small launch company that focuses on delivering small satellites to orbit.

The company's Electron rocket is designed for frequent, low-cost launches.

- Key Achievements:

 Rocket Lab has successfully launched dozens of missions, deploying small satellites for a variety of customers, including government agencies, commercial companies, and research institutions.

- Future Goals:
 Rocket Lab is developing a new rocket, Neutron, capable of carrying heavier payloads to orbit. The company also plans to expand its services to include missions to the Moon, Mars, and beyond.

6. Sierra Nevada Corporation (SNC)

- Overview:

 SNC, through its subsidiary Sierra Space, is developing the Dream Chaser spacecraft, which is designed to deliver cargo to the ISS and eventually carry crew.

- Key Achievements:

 Dream Chaser has been selected by NASA to provide cargo resupply missions to the ISS.

 The spacecraft's unique design allows it to land on conventional runways, offering flexibility in mission planning.

- Future Goals:

 SNC aims to develop a crewed version of Dream Chaser and expand its services to include deep space exploration and commercial space habitats.

7. Axiom Space

- Overview:

Axiom Space is working on developing the world's first commercial space station, which will initially be attached to the ISS before becoming a standalone platform.

- Key Achievements:

 Axiom has secured contracts to provide commercial modules to the ISS, and the company plans to conduct the first fully private crewed mission to the ISS in the near future.

- Future Goals:

 Axiom Space's ultimate goal is to create a commercial space station that will serve as a hub for research, manufacturing, and tourism in low Earth orbit.

8. Bigelow Aerospace

- Overview:

 Bigelow Aerospace is focused on developing expandable space habitats that can be used for a variety of purposes, including research, tourism, and even long-duration space missions.

- Key Achievements:

 The company successfully launched and tested its expandable habitat module, the BEAM, which is currently attached to the ISS.

- Future Goals:

 Bigelow Aerospace aims to create larger, standalone habitats that could serve as space stations or bases on the Moon or Mars.

The Role of International Cooperation

As the field of space exploration becomes increasingly global and collaborative, international cooperation is essential for overcoming the technical, financial, and logistical challenges of interplanetary travel.

Agencies like NASA, ESA, Roscosmos, and CNSA often collaborate on major missions, sharing resources, expertise, and technology.

This cooperation is exemplified by the International Space Station, a symbol of what can be achieved when nations work together.

In the future, international partnerships will likely be crucial for Mars exploration and beyond. The development of a global space economy, where resources from multiple countries and private entities are pooled, could accelerate the timeline for achieving human settlement on other planets.

Moreover, the establishment of international space laws and agreements will be necessary to ensure that space remains a domain of peaceful exploration, accessible to all humanity.

2. The Challenges of Space Travel

Having set the stage with an overview of current space travel efforts and the key players involved, we can now explore the numerous challenges that must be addressed to make interplanetary travel safe and feasible. This section will delve into the various obstacles—ranging from technical to physiological—that will need to be overcome as we venture further into the cosmos.

The Vastness of Space and Distance Between Planets

Space is unimaginably vast, with distances between planets that are difficult to comprehend.

For example, Mars, our most likely destination for human exploration, is approximately 225 million kilometers away from Earth on average.

Even at the fastest speeds achievable by current spacecraft, a one-way trip to Mars would take about six to nine months, depending on the alignment of the planets.

This extended travel time poses significant challenges, including the need for sufficient life support, protection from space hazards, and maintaining the physical and mental health of the crew.

- Travel Time:

 Reducing travel time is one of the most critical challenges in interplanetary travel.

 Current chemical propulsion systems are too slow for efficient travel to distant planets.

 As a result, research is focused on developing faster propulsion methods, such as nuclear propulsion or ion drives, which could significantly shorten the time needed to reach other planets.

- Communication Delays:

 The vast distances also result in communication delays. For instance, signals between Earth and Mars can take between 4 to 24 minutes to travel one way, depending on the relative positions of the planets.

 This delay complicates mission control and requires spacecraft to operate with a high degree of autonomy.

Radiation Exposure in Space

One of the most significant dangers of space travel is exposure to cosmic radiation. Outside Earth's protective atmosphere and magnetic field, astronauts are exposed to high levels of radiation from solar particles and galactic cosmic rays (GCRs). Prolonged exposure to this radiation can increase the risk of cancer, damage tissues and organs, and potentially lead to acute radiation sickness during solar storms.

- Radiation Sources:

The primary sources of harmful radiation in space include solar energetic particles (SEPs) emitted by the Sun during solar flares and coronal mass ejections, and GCRs, which are high-energy particles originating from outside our solar system.

- Current Mitigation Strategies:

  Current spacecraft are equipped with basic radiation shielding, primarily using materials like aluminum.

  However, these shields are not sufficient for long-duration missions.

  Advanced shielding techniques, including the use of water, polyethylene, or even magnetic or electric fields to deflect radiation, are being explored.

- Monitoring and Prediction:

 Effective radiation protection also requires real-time monitoring of space weather and predictive models to anticipate solar events that could pose a risk to astronauts.

Microgravity and Its Effects on Human Health

Microgravity, or the near-weightlessness experienced in space, has profound effects on the human body. Prolonged exposure to microgravity can lead to muscle atrophy, bone density loss, fluid redistribution, and changes in cardiovascular function. Astronauts who spend extended periods on the ISS undergo rigorous exercise regimens to mitigate these effects, but this is not a complete solution.

- Muscle and Bone Loss:

 In the absence of gravity, muscles do not need to work as hard to support the body, leading to atrophy.

 Similarly, bones lose density because they no longer bear weight.

 Over time, this can lead to an increased risk of fractures and other health issues.

- Fluid Redistribution:

 Microgravity causes bodily fluids to shift upwards, leading to facial puffiness, nasal congestion, and increased pressure in the head, which can affect vision.

- Countermeasures:

 Developing effective countermeasures for these health effects is crucial for long-duration missions.

 These may include advanced exercise equipment, medications to protect bone density, and artificial gravity solutions, such as rotating spacecraft sections to create centrifugal force.

Psychological Challenges of Long-duration Space Missions

Space travel is not only physically demanding but also psychologically challenging.

The isolation, confinement, and monotony of long-duration space missions can take a toll on the mental health of astronauts.

Moreover, the lack of natural light, the absence of Earthly stimuli, and the constant threat of life-threatening situations can exacerbate stress and anxiety.

- Isolation and Confinement: Astronauts on long-duration missions, such as those to Mars, will spend months or even years confined within a spacecraft, far from family, friends, and the familiar environment of Earth. The psychological impact of this isolation can lead to feelings of loneliness, depression, and anxiety. Moreover, the limited space aboard a spacecraft means that astronauts have little privacy and must live and work in close quarters with their crewmates, which can strain interpersonal relationships and lead to conflicts.

- Monotony and Sensory Deprivation:

The repetitive nature of daily routines and the lack of new experiences can contribute to a sense of monotony, which can be mentally draining. Additionally, the environment inside a spacecraft offers limited sensory stimulation—no fresh air, natural sounds, or changes in scenery. This sensory deprivation can affect cognitive function and emotional well-being, making it more challenging for astronauts to stay motivated and focused.

- Stress and Anxiety:

 The high-stakes nature of space missions, where every decision can have life-or-death consequences, adds to the psychological stress astronauts experience. They must be constantly vigilant, prepared to respond to emergencies, and capable of performing complex tasks under pressure. The awareness of being millions of kilometers from Earth, with limited support, can also exacerbate feelings of anxiety and fear.

- Psychological Support Systems:

 To address these challenges, space agencies are developing comprehensive psychological support systems.

 These include regular communication with psychologists on Earth, virtual reality environments that simulate Earth-like settings, and activities designed to promote mental well-being, such as exercise, creative outlets, and social interactions with crewmates.

Additionally, selecting astronauts who are psychologically resilient and have strong coping mechanisms is crucial for the success of long-duration missions.

Energy Requirements for Space Travel

The energy demands of space travel are immense, particularly for missions that involve traveling to other planets.

Spacecraft must carry enough fuel not only to reach their destination but also to power life support systems, communication equipment, and scientific instruments throughout the mission. For long-duration missions, like those to Mars or beyond, this poses significant challenges.

- Propulsion Energy:
 Current space missions rely primarily on chemical propulsion, which involves burning fuel to produce thrust.
 However, chemical rockets are not efficient for long-distance travel because they require large amounts of fuel, which adds weight and limits the payload capacity. To overcome this, researchers are exploring alternative propulsion systems, such as ion drives, nuclear propulsion, and solar sails, which offer higher efficiency and can potentially reduce travel time.

- Power Generation:

In addition to propulsion, spacecraft need a reliable source of power to operate onboard systems.

Solar panels are commonly used in space missions, but their effectiveness decreases with distance from the Sun, making them less viable for missions to the outer planets.

Nuclear power sources, such as radioisotope thermoelectric generators (RTGs), provide a more consistent energy supply and are already used in missions like the Voyager and Curiosity rovers.

However, advancements in energy storage and generation technologies will be necessary to meet the demands of future deep-space missions.

- Sustainability:

The sustainability of energy resources is a key consideration for long-duration missions. Developing closed-loop systems that can recycle and regenerate energy, such as advanced fuel cells or regenerative fuel systems, will be crucial. Additionally, in-situ resource utilization (ISRU) techniques, which involve harvesting energy from the environment of the destination planet (e.g., using solar energy on Mars), could provide a sustainable energy source for extended missions.

Challenges in Communication with Earth

Effective communication with Earth is essential for the success of space missions, but the vast distances involved in interplanetary travel pose significant challenges.

As spacecraft travel farther from Earth, communication delays increase, and the transmission of data becomes more complex.

- Communication Delays:

The speed of light sets a fundamental limit on how quickly signals can travel between Earth and a spacecraft.

For example, it takes about 4 to 24 minutes for a signal to travel between Earth and Mars, depending on their relative positions.

This delay makes real-time communication impossible and requires spacecraft to operate autonomously for extended periods.

- Data Transmission:

As missions venture farther into space, the amount of data that can be transmitted back to Earth decreases due to signal degradation and bandwidth limitations. High-gain antennas and deep space networks are used to mitigate these issues, but advancements in communication technology, such as laser-based communication systems, are needed to increase data transmission rates and ensure reliable communication over vast distances.

- Autonomy and AI:

 The communication delays necessitate that spacecraft be equipped with advanced autonomous systems capable of making decisions without immediate input from mission control. Artificial intelligence (AI) and machine learning algorithms are being developed to enable spacecraft to navigate, conduct scientific experiments, and respond to emergencies independently. These systems must be highly reliable and robust, given the critical nature of space missions.

3. Spacecraft Design and Engineering

With an understanding of the challenges posed by interplanetary travel, we can now turn our attention to the design and engineering of spacecraft that will enable humans to safely journey to other planets. This section will explore the current state of spacecraft technology, the necessary advancements in materials and systems, and the innovations required to overcome the challenges identified earlier.

Current Spacecraft Technology

Spacecraft have evolved significantly since the early days of space exploration, but the fundamental principles remain the same: they must protect their occupants from the harsh environment of space, provide life support, and enable propulsion and navigation.

- Space Shuttles:

NASA's Space Shuttle program, which operated from 1981 to 2011, was a pioneering effort in reusable spacecraft.

The Space Shuttle was capable of carrying astronauts and cargo to low Earth orbit (LEO) and returning them safely to Earth.

However, the program was expensive and ultimately unsustainable for long-term space exploration.

- The International Space Station (ISS):

 The ISS is a testament to international collaboration and human ingenuity. Orbiting approximately 400 kilometers above Earth, the ISS serves as a microgravity laboratory and a testbed for technologies that will be essential for future deep-space missions.

 The ISS has provided invaluable insights into long-duration spaceflight and the effects of microgravity on the human body.

- Crewed Capsules:

  Modern crewed spacecraft, such as SpaceX's Crew Dragon and Boeing's CST-100 Starliner, represent the next generation of space vehicles.

  These capsules are designed for transporting astronauts to and from the ISS, but their technology could be adapted for deep-space missions.

  They feature advanced life support systems, heat shields for re-entry, and the capability to dock with space stations.

Necessary Advancements in Spacecraft Materials

The materials used in spacecraft construction must be lightweight, durable, and capable of withstanding the extreme conditions of space, including temperature fluctuations, radiation, and micrometeoroid impacts.

As we look toward interplanetary travel, new materials will be required to meet these challenges.

- Radiation Shielding Materials:

 Protecting astronauts from cosmic radiation is one of the biggest challenges in spacecraft design.

 Traditional materials, like aluminum, provide some protection but are not sufficient for long-duration missions.

Researchers are exploring advanced materials, such as polyethylene, which is rich in hydrogen and effective at blocking radiation.

Additionally, concepts like multi-layered shields with embedded hydrogen-rich materials or even water jackets around the crew habitat are being investigated.

- Lightweight Composites:

Reducing the weight of spacecraft is crucial for improving fuel efficiency and increasing payload capacity.

Advanced composites, such as carbon fiber reinforced polymers (CFRPs), offer high strength-to-weight ratios and are increasingly used in spacecraft construction.

These materials are not only lighter but also more durable than traditional metals, making them ideal for the harsh conditions of space.

- Self-Healing Materials:

 To address the risk of micrometeoroid impacts and other damage, researchers are developing self-healing materials that can repair themselves when punctured or cracked.

 These materials could significantly enhance the safety and longevity of spacecraft, particularly on long-duration missions where repairs may be difficult.

Propulsion Systems for Long-Distance Travel

Propulsion systems are the heart of any spacecraft, determining how quickly and efficiently it can travel through space.

The limitations of current propulsion technologies are a major barrier to interplanetary travel, prompting the development of new and more advanced systems.

- Chemical Propulsion:

 Traditional chemical rockets, which burn fuel to produce thrust, are powerful but inefficient for long-distance travel.

 They require large amounts of fuel, which increases the weight and cost of the mission. Despite these limitations, chemical propulsion is still used for launches and short-distance missions, as it provides the high thrust needed to escape Earth's gravity.

- Ion Drives:
 Ion propulsion systems offer a more efficient alternative to chemical rockets. By accelerating ions (charged particles) to extremely high speeds, these drives can generate continuous thrust over long periods, gradually increasing the spacecraft's velocity. NASA's Dawn mission, which explored the asteroid belt, successfully used ion propulsion to visit multiple celestial bodies. While ion drives are slower to accelerate, their efficiency makes them ideal for deep-space missions.

- Nuclear Propulsion:

Nuclear thermal propulsion (NTP) and nuclear electric propulsion (NEP) are two promising technologies for interplanetary travel.

NTP uses a nuclear reactor to heat a propellant, such as hydrogen, which is then expelled to produce thrust.

NEP, on the other hand, uses a nuclear reactor to generate electricity, which powers an ion drive or other electric propulsion system.

Both methods offer higher efficiency than chemical propulsion and could significantly reduce travel time to Mars and beyond.

- Fusion Propulsion:

 Although still in the experimental stage, fusion propulsion holds the potential to revolutionize space travel.

 By harnessing the energy released from nuclear fusion—the same process that powers the Sun—these systems could provide unprecedented levels of thrust and efficiency.

 Fusion propulsion could enable faster travel to the outer planets and even interstellar exploration.

- Solar Sails and Light-based Propulsion:

 Solar sails use the pressure of sunlight to propel a spacecraft, offering a fuel-free method of propulsion. While the acceleration is slow, solar sails can reach high speeds over time, making them suitable for missions to distant planets or even other star systems. Light-based propulsion concepts, such as those involving laser beams, could further enhance the capabilities of solar sails, enabling rapid travel across vast distances.

Life Support Systems and Sustainability

Ensuring the survival and well-being of astronauts on long-duration missions requires advanced life support systems that can provide air, water, food, and waste management for extended periods.

These systems must be reliable, efficient, and capable of functioning autonomously.

- Oxygen Production and Carbon Dioxide Removal:

 Oxygen is essential for human survival, and spacecraft must have systems in place to produce and recycle it.

 Current systems, like those on the ISS, use electrolysis to split water into oxygen and hydrogen.

Carbon dioxide removal is equally important, as high levels of CO_2 can be toxic.

Technologies such as molecular sieves and chemical scrubbers are used to capture and remove CO_2 from the cabin air.

Future systems may incorporate biological processes, such as photosynthesis by algae, to produce oxygen and remove CO_2 simultaneously.

- Water Recycling and Management:

Water is a precious resource in space, and efficient recycling systems are essential for long-duration missions.

The ISS uses a Water Recovery System (WRS) that recycles urine, sweat, and humidity into potable water.

However, more advanced systems will be needed for missions to Mars and beyond, where resupply from Earth is not an option.

Future technologies may include more efficient filtration and purification systems, as well as methods for extracting water from local resources, such as Martian soil.

- Food Production and Nutrition:

 Providing a stable, nutritious food supply for long-duration missions is a major challenge.

 Current space missions rely on pre-packaged meals, but these have limitations in terms of shelf life and variety.

Researchers are exploring methods for growing food in space, such as hydroponics and aeroponics, which allow plants to grow without soil.

Advances in biotechnology could also enable the production of lab-grown meat and other protein sources.

Ensuring a balanced diet that meets the nutritional needs of astronauts is critical for maintaining their health and performance.

- Waste Management and Recycling Systems:

 Efficient waste management is essential for maintaining a healthy environment aboard a spacecraft.

 Current systems involve compacting and storing waste, which is then disposed of when the spacecraft returns to Earth.

For long-duration missions, however, this approach is not feasible.

Instead, spacecraft will need to be equipped with recycling systems that can convert waste into usable resources, such as water, nutrients for plants, and even fuel.

Advanced waste processing technologies, including microbial systems that break down waste into basic components, are being developed to meet these needs.

Artificial Gravity Solutions

 One of the most significant challenges of long-duration space travel is the impact of microgravity on the human body.

 Prolonged exposure to microgravity can lead to muscle atrophy, bone density loss, and other health issues, as mentioned earlier. Creating artificial gravity within a spacecraft could mitigate these effects and improve the overall health and well-being of astronauts.

- Rotating Habitats:

 The most widely studied method for creating artificial gravity is through the use of rotating habitats.

 By spinning a section of the spacecraft, centrifugal force can simulate the effects of gravity.

The speed of rotation and the radius of the habitat determine the strength of the artificial gravity.

One challenge with this approach is managing the Coriolis effect, which can cause disorientation and motion sickness.

Designing a habitat with a large enough radius to reduce these effects is key to the success of this method.

- Tether Systems:

 Another concept involves using tethers to connect two sections of a spacecraft, which are then spun around a common center of mass.

 This creates artificial gravity in both sections. Tether systems are relatively simple and lightweight, but they require precise control to maintain stability and prevent the tether from tangling or breaking.

- Hybrid Systems:

 Hybrid systems that combine rotating habitats with traditional microgravity environments could offer a flexible solution.

 For example, a spacecraft might include a rotating section for exercise and sleep, while other activities take place in microgravity. This approach allows astronauts to experience the benefits of artificial gravity while still taking advantage of the unique conditions of microgravity for scientific research.

Thermal Control Systems

Spacecraft must be equipped with thermal control systems to regulate temperature and prevent overheating or freezing.

The vacuum of space provides no medium for heat to be transferred by convection, so spacecraft must rely on radiation and conduction to manage heat.

- Radiators and Heat Pipes:

Radiators are used to dissipate heat generated by the spacecraft's systems and equipment.

Heat pipes, which transfer heat from one part of the spacecraft to another, are often used in conjunction with radiators.

These systems must be carefully designed to ensure that heat is distributed evenly and that critical components do not overheat or become too cold.

- Insulation and Thermal Blankets:

 Insulation materials, such as multi-layer insulation (MLI), are used to minimize heat loss or gain from the spacecraft's external environment.

 These thermal blankets consist of multiple layers of thin, reflective material, which trap heat and prevent it from radiating away too quickly.

They also protect the spacecraft from extreme temperature fluctuations, such as those encountered when moving from the shadow of a planet into direct sunlight.

Insulation is crucial for maintaining the internal temperature of the spacecraft within a range that is safe for both the crew and the onboard equipment.

- Active Thermal Control Systems:
 In addition to passive systems like radiators and insulation, spacecraft also rely on active thermal control systems to manage heat. These systems include pumps, fans, and fluid loops that circulate coolant to remove excess heat from high-temperature components, such as electronics and propulsion systems. Active thermal control is especially important for managing the heat generated by energy-intensive activities, like firing propulsion engines or operating scientific instruments.

4. Radiation Protection

Radiation exposure is one of the most significant hazards for astronauts on long-duration space missions, especially when traveling beyond Earth's protective magnetosphere.

Effective radiation protection is essential for ensuring the safety and health of crew members as they journey to other planets.

This section will explore the sources of space radiation, its effects on human health, and the innovative materials and technologies being developed to shield astronauts from harmful radiation.

Sources of Space Radiation

Space radiation comes from a variety of sources, each with different characteristics and risks.

The two primary sources of radiation that pose a threat to astronauts are solar energetic particles (SEPs) and galactic cosmic rays (GCRs).

- Solar Energetic Particles (SEPs):

SEPs are high-energy particles emitted by the Sun, particularly during solar flares and coronal mass ejections (CMEs). These particles include protons, electrons, and heavier ions, and they can reach Earth within minutes to hours of a solar event. SEPs are particularly dangerous because they can deliver a sudden, intense dose of radiation, potentially overwhelming shielding and posing an immediate risk to astronauts.

- Galactic Cosmic Rays (GCRs):

GCRs are high-energy particles that originate outside our solar system, likely from supernova explosions and other cosmic events. They consist mostly of protons, but also include heavier nuclei, such as helium and iron. GCRs are highly penetrating and can travel through spacecraft and human tissue, causing damage at the cellular and DNA levels. Unlike SEPs, GCRs are a constant, low-level threat that must be mitigated over the entire duration of a space mission.

- Trapped Radiation Belts:

In addition to SEPs and GCRs, astronauts near Earth must also contend with the Van Allen radiation belts, which are regions of trapped charged particles surrounding the planet. These belts are not a concern for deep-space missions, but they pose a significant risk during launches and in low Earth orbit (LEO). Spacecraft must be designed to minimize exposure to these belts, especially during crewed missions to the Moon and beyond.

Effects of Radiation on Human Health

Prolonged exposure to space radiation can have severe consequences for human health.

Understanding these risks is crucial for developing effective countermeasures.

Acute Radiation Sickness:

 In the event of a significant radiation event, such as a solar flare, astronauts could be exposed to a high dose of radiation in a short period.

 This could lead to acute radiation sickness, characterized by symptoms such as nausea, vomiting, fatigue, and a decrease in white blood cells, which weakens the immune system. In extreme cases, acute radiation exposure can be fatal.

- Increased Cancer Risk:

 One of the most concerning long-term effects of radiation exposure is an increased risk of cancer.

 Radiation can damage DNA, leading to mutations that may cause cancer to develop years or even decades after exposure.

 The risk of cancer is particularly high from exposure to GCRs, which can penetrate deep into tissues and cause widespread cellular damage.

- Cognitive and Neurological Effects:
Recent studies have suggested that exposure to space radiation, particularly GCRs, may have adverse effects on cognitive function and neurological health. These effects could include memory loss, impaired decision-making, and an increased risk of neurodegenerative diseases such as Alzheimer's. The exact mechanisms behind these effects are not fully understood, but they represent a significant concern for long-duration space missions.

- Cardiovascular Disease:

 Radiation exposure has also been linked to an increased risk of cardiovascular disease.

 Damage to the cells lining blood vessels can lead to the development of atherosclerosis (hardening of the arteries), which increases the risk of heart attacks and strokes.

 This risk must be carefully managed in astronauts who will be exposed to elevated levels of radiation for extended periods.

Current Radiation Shielding Techniques

Shielding is the primary method for protecting astronauts from space radiation.

However, the effectiveness of shielding depends on the materials used and the type of radiation being blocked.

- Traditional Shielding Materials:

Aluminum is commonly used in spacecraft construction due to its light weight and relatively good shielding properties.

However, aluminum alone is not sufficient to protect against all types of space radiation, especially GCRs. Other materials, such as lead, are more effective at blocking radiation but are too heavy to be practical for space missions.

- Hydrogen-rich Materials:

 Materials rich in hydrogen, such as polyethylene, are more effective at blocking GCRs because hydrogen atoms are effective at absorbing and scattering incoming radiation.

 Polyethylene is currently used in some spacecraft and spacesuits as a lightweight, effective shield against radiation. Researchers are also exploring other hydrogen-rich compounds that could provide even better protection.

- Water as Shielding:

Water is another excellent radiation shield due to its high hydrogen content. Some spacecraft designs propose using water stored in tanks around the crew habitat to provide radiation protection. This approach has the added benefit of making use of a resource that is already needed for life support, thus maximizing efficiency. However, storing large amounts of water adds weight to the spacecraft, which must be carefully managed.

Innovative Materials and Technologies for Radiation Protection

To improve the effectiveness of radiation shielding for deep-space missions, researchers are exploring new materials and technologies that could provide better protection while minimizing weight and bulk.

- Multi-layered Shielding:

 One approach to improving radiation protection is the use of multi-layered shielding, which combines different materials to block a broader range of radiation types.

 For example, a layered shield might include a hydrogen-rich material to absorb GCRs, followed by a denser material like tungsten to block SEPs. This combination can provide more comprehensive protection than any single material.

- Electromagnetic Shields:

 Some researchers are investigating the use of electromagnetic fields to create a protective shield around the spacecraft, similar to Earth's magnetic field.

 These shields could potentially deflect incoming charged particles, reducing the radiation dose experienced by the crew. While this concept is still in the experimental stage, it holds promise as a lightweight, effective solution for radiation protection.

- Radiation-Resistant Polymers:

 Advances in polymer science have led to the development of new materials that are more resistant to radiation damage.

 These polymers can be used in spacesuits, spacecraft interiors, and other critical areas to provide additional protection for astronauts.

 Some of these materials can also repair themselves when exposed to radiation, extending their lifespan and effectiveness.

- Biological Countermeasures:

In addition to physical shielding, researchers are exploring biological countermeasures to mitigate the effects of radiation on the human body. These include drugs and supplements that can enhance DNA repair mechanisms, reduce inflammation, and protect cells from oxidative damage. For example, antioxidants such as vitamins C and E are being studied for their potential to reduce radiation-induced damage.

Strategies for Minimizing Exposure During Space Travel

In addition to improving shielding, mission planners must develop strategies to minimize radiation exposure during space travel.

These strategies include careful timing of missions, optimizing spacecraft design, and establishing safe havens for astronauts.

- Mission Timing:

  Solar activity follows an 11-year cycle, with periods of high and low activity.

  Planning missions during periods of low solar activity can reduce the risk of exposure to SEPs.

  Additionally, missions can be timed to avoid solar storms or other space weather events, which can increase radiation levels.

- Optimized Spacecraft Design:

 The layout and design of the spacecraft can be optimized to minimize radiation exposure.

 For example, critical systems and crew quarters can be placed in areas that are better shielded, while less sensitive equipment is positioned in areas with less protection.
 Additionally, the use of modular spacecraft designs can allow for the rearrangement of components to create more effective shielding configurations.

- Safe Havens:
 Spacecraft can be equipped with "safe havens"—areas that provide enhanced radiation protection where astronauts can take shelter during solar storms or other radiation events. These safe havens could be reinforced with additional shielding materials or water tanks to provide maximum protection. The safe haven concept is particularly important for missions to Mars, where astronauts will be exposed to radiation for extended periods without the possibility of returning to Earth.

5. Life Support Systems and Habitability

Ensuring the survival and well-being of astronauts on long-duration missions requires advanced life support systems that can sustain human life in the harsh environment of space.

These systems must provide oxygen, water, food, and waste management while maintaining a comfortable and safe living environment.

This section explores the challenges and innovations in life support systems and the design of habitats that will allow humans to live and thrive in space for extended periods.

Oxygen Production and Carbon Dioxide Removal Systems

Oxygen is vital for human survival, and spacecraft must be equipped with reliable systems to produce and maintain an adequate supply.

Additionally, the removal of carbon dioxide (CO_2) is essential to prevent it from reaching toxic levels.

- Oxygen Generation:

The primary method for generating oxygen in spacecraft is through the electrolysis of water.

This process uses electricity to split water molecules into hydrogen and oxygen.

The oxygen is then delivered to the crew, while the hydrogen can be vented into space or used as a propellant.

On the ISS, the Oxygen Generation Assembly (OGA) uses this method to provide breathable air for astronauts.

For long-duration missions, advancements in electrolysis efficiency and reliability will be crucial.

- Carbon Dioxide Removal:
 CO_2 must be continuously removed from the spacecraft's atmosphere to prevent it from accumulating to dangerous levels. The ISS uses the Carbon Dioxide Removal Assembly (CDRA), which employs molecular sieves to absorb CO_2 from the air. The absorbed CO_2 is then vented into space. Future systems may incorporate more efficient and compact designs, as well as regenerative systems that can convert CO_2 back into oxygen through processes such as photosynthesis or chemical reactions.

- Bioregenerative Life Support Systems:

An emerging area of research is the development of bioregenerative life support systems, which use living organisms to produce oxygen, remove CO_2, and even recycle waste.

For example, algae or plants could be grown onboard the spacecraft to produce oxygen and absorb CO_2 through photosynthesis.

These systems offer the potential for a closed-loop life support system, reducing the need for resupply missions and increasing the sustainability of long-duration space travel.

Water Recycling and Management

Water is a critical resource for astronauts, used for drinking, hygiene, food preparation, and oxygen production.

Given the constraints of space travel, efficient water recycling and management systems are essential.

- Water Recovery Systems:

The ISS is equipped with a Water Recovery System (WRS) that recycles urine, sweat, and humidity from the air into potable water. This system relies on filtration, distillation, and chemical treatment to purify the water and make it safe for consumption. For future missions, improving the efficiency and reliability of these systems will be important, as will the development of compact, lightweight designs that can be easily integrated into spacecraft.

- In-Situ Resource Utilization (ISRU):

 For missions to the Moon or Mars, water can potentially be sourced from local resources, such as ice deposits or hydrated minerals. ISRU technologies are being developed to extract and purify this water, which can then be used for drinking, agriculture, or even fuel production. The ability to harvest water from the destination planet or moon would greatly reduce the need for water storage and resupply from Earth.

- Advanced Water Purification:

Researchers are exploring new methods for purifying water in space, including advanced filtration membranes, ultraviolet (UV) sterilization, and electrochemical treatment.

These technologies could improve the efficiency and effectiveness of water recycling systems, ensuring a reliable supply of clean water for astronauts on long-duration missions.

Food Production and Nutrition in Space

Providing a stable, nutritious food supply is essential for maintaining the health and performance of astronauts on long-duration missions.

Current space missions rely on pre-packaged meals, but this approach has limitations in terms of variety, freshness, and nutritional content.

- Pre-packaged Meals:
 Pre-packaged space food is designed to be lightweight, shelf-stable, and easy to prepare. However, the nutritional content of these meals can degrade over time, especially during extended missions. Ensuring that astronauts receive a balanced diet with adequate vitamins, minerals, and calories is a major challenge. To address this, space agencies are working on improving the formulation and preservation of space food, including the development of high-nutrient, long-shelf-life products.

- Growing Food in Space:

To supplement pre-packaged meals and provide fresh produce, researchers are exploring methods for growing food in space.

Hydroponics, which involves growing plants in a nutrient-rich water solution, and aeroponics, where plants are grown in an air or mist environment, are two promising techniques.

These methods do not require soil and can be adapted to the microgravity environment of space.

Experiments on the ISS, such as the Veggie project, have successfully grown lettuce, radishes, and other crops, demonstrating the feasibility of space agriculture.

- Biotechnology and Lab-grown Food:

 Advances in biotechnology could enable the production of lab-grown meat and other protein sources in space.

 Cultured meat, which is produced by growing animal cells in a controlled environment, offers the potential for fresh, high-protein food without the need for livestock.

Additionally, microbial fermentation could be used to produce a variety of food products, including proteins, vitamins, and flavorings, from simple inputs like sugar and water.

- Ensuring Nutritional Balance:

 Space agencies must carefully plan astronauts' diets to ensure they receive the necessary nutrients for long-term health. This includes accounting for the increased nutritional needs due to the physical and psychological demands of space travel. Supplements and fortified foods may be used to address any nutritional gaps, and ongoing research will continue to refine our understanding of human nutritional requirements in space.

Waste Management and Recycling Systems

Efficient waste management is critical for maintaining a healthy and safe environment aboard a spacecraft.

Waste must be carefully managed to prevent contamination, odors, and the buildup of harmful gases.

- Solid Waste Management:

 Solid waste, including food packaging, personal hygiene products, and human waste, must be safely stored and processed.

 On the ISS, solid waste is compacted and stored in containers that are later disposed of by burning up in the atmosphere during re-entry aboard cargo spacecraft.

For long-duration missions, however, new approaches will be needed.

Researchers are exploring technologies that can convert solid waste into usable resources, such as water, nutrients, or even fuel, through processes like pyrolysis (thermal decomposition) and microbial digestion.

- Liquid Waste Recycling:

 Liquid waste, primarily urine, is already recycled on the ISS through the Water Recovery System.

 This process involves filtering and purifying the urine to produce clean, potable water.

 Further improvements in efficiency and reliability will be essential for future missions, as well as the development of systems that can handle other liquid wastes, such as greywater from hygiene activities.

- Air and Atmosphere Management:
Waste management also includes controlling the spacecraft's atmosphere to maintain safe levels of oxygen, carbon dioxide, and other gases. Trace contaminants, such as volatile organic compounds (VOCs) released from materials and equipment, must be removed using activated carbon filters or other air purification technologies. Maintaining the balance of gases in the spacecraft's atmosphere is critical for the health and well-being of the crew.

Habitat Design for Comfort and Mental Well-being

The design of spacecraft habitats plays a crucial role in ensuring the comfort, safety, and mental well-being of astronauts on long-duration missions.

A well-designed habitat can help mitigate the psychological challenges of isolation and confinement, promote social interaction, and provide a sense of connection to Earth.

- Living Space:

 Spacecraft habitats must provide enough living space for the crew to move freely, perform daily activities, and have privacy when needed.
 The layout of the habitat should be carefully planned to maximize the use of available space, while also allowing for personal space where astronauts can retreat and relax. Modular designs that allow for reconfiguration of the living area could provide flexibility and help reduce feelings of confinement.

- Lighting and Circadian Rhythms:

 Proper lighting is essential for maintaining astronauts' circadian rhythms, which regulate sleep and wake cycles.

 Exposure to natural light is crucial for synchronizing these rhythms, but this is difficult to achieve in space.

LED lighting systems that can simulate the natural progression of daylight are being developed to help astronauts maintain a healthy sleep schedule.

Additionally, windows that provide views of Earth and space can offer psychological benefits by connecting astronauts to their home planet.

- Recreational and Social Spaces:
 Providing spaces for recreation and social interaction is important for maintaining crew morale and mental health. Recreational activities, such as exercise, games, and creative outlets like music or art, can help alleviate stress and boredom. Social spaces where the crew can gather for meals, discussions, or relaxation can foster a sense of camaraderie and teamwork, which is critical for maintaining positive interpersonal relationships during long missions.

- Mental Health Support:

 Long-duration missions pose significant psychological challenges, including isolation, confinement, and the stress of being far from home.

 Spacecraft habitats should be designed with mental health in mind, providing spaces for relaxation, privacy, and social interaction.

Additionally, access to psychological support, both through communication with Earth-based professionals and through onboard resources like virtual reality environments, can help astronauts manage stress and maintain their mental well-being.

Psychological Support Systems for Astronauts

The mental health of astronauts is as important as their physical health, especially during long-duration missions where isolation and confinement can take a toll on their well-being.

Psychological support systems must be integrated into the design of the spacecraft and the mission plan to ensure that astronauts remain mentally healthy and resilient.

- Pre-mission Training:

 Preparing astronauts for the psychological challenges of space travel begins long before they leave Earth.

 Pre-mission training includes simulations of space conditions, team-building exercises, and psychological assessments to identify potential stressors and coping strategies. Astronauts are also trained in self-care techniques, such as mindfulness and stress management, which they can use during the mission.

- In-mission Support:

 During the mission, regular communication with psychologists and mental health professionals on Earth is crucial. Video calls, emails, and other forms of communication can provide support and guidance to astronauts as they navigate the challenges of space travel. In addition, virtual reality (VR) technology is being explored as a tool for providing immersive, Earth-like experiences, such as virtual nature walks or social interactions, which can help mitigate feelings of isolation.

- Post-mission Care:

 After returning to Earth, astronauts may experience challenges readjusting to life on the planet.
 Post-mission care includes debriefings, physical rehabilitation, and psychological support to help them transition back to normal life.

 Understanding the long-term psychological effects of space travel is an ongoing area of research, with the goal of improving support systems for future missions.

6. Propulsion Systems for Interplanetary Travel

The development of advanced propulsion systems is critical for enabling safe and efficient interplanetary travel.

Current propulsion technologies are not sufficient for reaching distant planets within reasonable timeframes, so new and innovative solutions are needed.

This section explores the various propulsion systems under development, their potential applications, and the challenges that must be overcome to make interplanetary travel a reality.

Overview of Existing Propulsion Systems

- Chemical Propulsion:

 Chemical rockets have been the backbone of space travel since the dawn of the space age.

 These rockets work by burning a propellant (fuel and oxidizer) to produce a high-speed exhaust, which generates thrust.

While chemical propulsion provides high thrust, it is not very efficient in terms of fuel consumption, which limits the range and payload capacity of missions.

Chemical rockets are best suited for launches from Earth and short-duration missions within the inner solar system.

- Electric Propulsion:
 Electric propulsion systems, such as ion drives, offer a more efficient alternative to chemical rockets. These systems use electricity to ionize a propellant (usually xenon) and accelerate the ions to produce thrust. Ion drives generate much less thrust than chemical rockets, but they can operate continuously for long periods, gradually increasing the spacecraft's velocity. Electric propulsion is ideal for deep-space missions, where efficiency is more important than rapid acceleration.

- Nuclear Propulsion:

 Nuclear propulsion is a promising technology for interplanetary travel, offering higher efficiency and greater thrust than chemical and electric propulsion systems.

 There are two main types of nuclear propulsion: nuclear thermal propulsion (NTP) and nuclear electric propulsion (NEP).

- Nuclear Thermal Propulsion (NTP):

NTP systems use a nuclear reactor to heat a propellant, such as hydrogen, to high temperatures. The hot gas is then expelled through a nozzle to produce thrust. NTP offers a significant improvement in efficiency over chemical rockets and could reduce travel time to Mars by up to 50%. However, the development of NTP technology faces challenges related to reactor safety, radiation shielding, and fuel availability.

- Nuclear Electric Propulsion (NEP):

NEP systems use a nuclear reactor to generate electricity, which is then used to power an electric propulsion system, such as an ion drive.

NEP offers even greater efficiency than NTP, making it well-suited for long-duration missions to the outer planets. However, NEP systems produce lower thrust than NTP, so they are best suited for missions that do not require rapid acceleration.

- Solar Sails and Light-based Propulsion:

 Solar sails use the pressure of sunlight to propel a spacecraft.

 These large, reflective sails can gradually accelerate a spacecraft to high speeds without the need for propellant.

 Solar sails are most effective for missions within the inner solar system, where sunlight is abundant.

Light-based propulsion, which uses powerful lasers to push a spacecraft equipped with a reflective sail, could enable even faster travel, potentially allowing for interstellar missions.

However, these technologies are still in the experimental stage and face significant engineering challenges.

- Fusion Propulsion:

 Fusion propulsion holds the potential to revolutionize space travel by providing a nearly limitless source of energy.

 Fusion drives would work by fusing light atomic nuclei, such as hydrogen, to release vast amounts of energy, which could then be used to generate thrust.

While fusion propulsion could enable rapid travel to the outer planets and beyond, the technology is still in its infancy, with significant technical hurdles to overcome, including achieving and sustaining controlled fusion reactions in a compact, lightweight system suitable for spacecraft.

- Antimatter Propulsion:
Antimatter propulsion is a theoretical concept that could provide the ultimate in propulsion efficiency. When matter and antimatter come into contact, they annihilate each other, releasing an enormous amount of energy. If this energy could be harnessed for propulsion, it could enable extremely fast travel across the solar system or even to other star systems. However, antimatter is extremely difficult and expensive to produce and store, making this technology purely speculative at present.

Nuclear Propulsion and Its Potential

Nuclear propulsion offers one of the most promising paths forward for interplanetary travel, combining high efficiency with substantial thrust.

There are several reasons why nuclear propulsion is seen as a key technology for future space missions.

- Reduced Travel Time:

 One of the main advantages of nuclear propulsion is its ability to significantly reduce travel time to other planets.

 For example, a mission to Mars using nuclear thermal propulsion could take as little as four to six months, compared to six to nine months with chemical propulsion. Shorter travel times reduce the time astronauts are exposed to space radiation and other hazards, improving the overall safety of the mission.

- Higher Efficiency:
Nuclear propulsion systems have a much higher specific impulse (a measure of propulsion efficiency) than chemical rockets. This means that they require less fuel to achieve the same velocity, allowing spacecraft to carry more cargo or travel further with the same amount of fuel. This increased efficiency is crucial for long-duration missions, where every kilogram of fuel saved can be used for other purposes, such as scientific instruments or life support systems.

- Challenges and Risks:
 Despite its potential, nuclear propulsion faces significant challenges. Developing a safe and reliable nuclear reactor for space use is a complex task, requiring advanced materials, radiation shielding, and fail-safe mechanisms to prevent accidents. Additionally, there are political and environmental concerns associated with launching nuclear reactors into space, particularly in the event of a launch failure. Addressing these challenges will require ongoing research and international cooperation.

Fusion and Antimatter Propulsion (Theoretical Concepts)

While nuclear propulsion is within reach, fusion and antimatter propulsion remain largely theoretical, offering tantalizing possibilities for the future of space travel.

- Fusion Propulsion:

Fusion propulsion is based on the same principles that power the Sun, where light atomic nuclei combine to form heavier nuclei, releasing vast amounts of energy.

If harnessed for space travel, fusion propulsion could provide a virtually limitless source of energy, enabling rapid travel to distant planets and even other star systems.

Current research is focused on developing compact, stable fusion reactors that can be used in space.

Challenges include achieving and sustaining the high temperatures and pressures needed for fusion, as well as managing the radiation produced by the reaction.

- Antimatter Propulsion:

 Antimatter propulsion is perhaps the most speculative of all propulsion concepts, relying on the annihilation of matter and antimatter to produce thrust.

 When matter and antimatter collide, they are converted entirely into energy, providing an extremely efficient form of propulsion.

However, antimatter is incredibly difficult to produce and store, and the quantities required for space travel are far beyond current capabilities.

While antimatter propulsion could potentially enable travel to other star systems, it remains a distant possibility that will require breakthroughs in both physics and engineering.

Solar Sails and Light-based Propulsion

Solar sails and light-based propulsion represent innovative approaches to space travel that do not rely on traditional propellants.

These technologies offer unique advantages for certain types of missions, particularly those involving long-duration travel in the inner solar system or even interstellar exploration.

- Solar Sails:

 Solar sails work by capturing the momentum of photons from sunlight. Although the force exerted by sunlight is very small, it is continuous, allowing a spacecraft equipped with a solar sail to gradually build up speed over time. Solar sails are particularly well-suited for missions that require long-duration travel without the need for significant onboard fuel. They have been successfully demonstrated in missions such as JAXA's IKAROS and NASA's LightSail.

- Light-based Propulsion:

Light-based propulsion involves using powerful lasers to push a spacecraft equipped with a reflective sail.

This concept, sometimes referred to as "laser sailing," could potentially enable rapid travel within the solar system or even to nearby star systems.

Projects like Breakthrough Starshot are exploring the feasibility of sending tiny, lightweight spacecraft to other star systems using this technology.

However, light-based propulsion faces significant engineering challenges, including the development of extremely powerful and precise lasers, as well as the ability to steer and decelerate the spacecraft once it reaches its destination.

Energy Requirements and Fuel Efficiency

One of the critical factors in the success of interplanetary travel is the management of energy resources and fuel efficiency.

The choice of propulsion system, the design of the spacecraft, and the mission profile all play a role in determining the energy requirements for a given mission.

- Specific Impulse:

 Specific impulse (Isp) is a key metric used to evaluate the efficiency of a propulsion system. It measures the amount of thrust produced per unit of propellant consumed. Higher specific impulse values indicate more efficient propulsion systems, which require less fuel for the same amount of thrust. Electric propulsion systems, such as ion drives, typically have much higher specific impulses than chemical rockets, making them more suitable for long-duration missions.

- Energy Sources:

The energy required for propulsion can come from various sources, including chemical reactions, nuclear fission or fusion, solar power, or even antimatter.

Each energy source has its advantages and limitations. For example, chemical propulsion is well understood and capable of producing high thrust, but it is limited by the availability of propellant.

Nuclear propulsion offers higher efficiency and greater energy density, but it poses challenges related to safety and radiation.

Solar power is abundant but less effective in the outer solar system, where sunlight is weaker.

- Fuel Efficiency:

Fuel efficiency is critical for reducing the overall mass of the spacecraft, which in turn reduces the energy required for launch and travel. Efficient propulsion systems allow for longer missions, larger payloads, and more flexibility in mission planning. Innovations in propulsion technology, such as the development of more efficient engines and the use of alternative propellants, are key to improving fuel efficiency and making interplanetary travel more feasible.

Future Research and Development in Propulsion Technologies

The future of interplanetary travel depends on continued research and development in propulsion technologies.

Several areas of research hold particular promise for advancing our capabilities.

- Advanced Propulsion Systems:

Research into advanced propulsion systems, such as nuclear thermal and electric propulsion, is critical for enabling missions to Mars and beyond.

These systems offer significant improvements in efficiency and thrust compared to current chemical rockets, making them well-suited for deep-space exploration. Ongoing research is focused on improving the safety, reliability, and performance of these systems.

- Hybrid Propulsion:
 Hybrid propulsion systems, which combine different types of propulsion, could offer greater flexibility and efficiency for interplanetary missions.

 For example, a spacecraft could use chemical propulsion for launch and orbital insertion, followed by nuclear or electric propulsion for the cruise phase of the mission. Hybrid systems could optimize fuel use and reduce travel time, making them a promising area of research.

- In-Situ Propellant Production:

 The ability to produce propellant at the destination planet or moon could greatly reduce the amount of fuel that needs to be carried from Earth.

 In-situ resource utilization (ISRU) techniques, such as extracting water from Martian ice and converting it into hydrogen and oxygen for fuel, are being explored as a way to make missions more sustainable and cost-effective.

- Breakthrough Technologies: The pursuit of breakthrough technologies, such as fusion propulsion or antimatter drives, represents the long-term future of space travel. While these technologies are still in the experimental or theoretical stages, they have the potential to revolutionize our ability to explore the solar system and beyond. Continued investment in fundamental research, as well as collaboration between government, industry, and academia, will be essential for making these technologies a reality.

7. Artificial Gravity and Human Health

Artificial gravity is a critical consideration for long-duration space missions, as it can help mitigate the adverse health effects of prolonged exposure to microgravity.

Creating artificial gravity within a spacecraft requires innovative engineering solutions, and its impact on human physiology is a key area of research.

This section explores the importance of gravity for human health, the concepts of artificial gravity, and the challenges of implementing it in space travel.

Importance of Gravity for Human Health

Gravity plays a crucial role in maintaining human health, particularly in the areas of bone density, muscle mass, cardiovascular function, and overall physical well-being.

The absence of gravity, as experienced in microgravity environments like the International Space Station (ISS), can lead to a range of health issues.

- Bone Density Loss:

In the absence of gravity, bones are not subjected to the mechanical stress that typically stimulates bone formation. As a result, astronauts in microgravity environments experience a reduction in bone density, particularly in weight-bearing bones such as the spine, hips, and legs. This condition, known as spaceflight osteopenia, increases the risk of fractures and may have long-term consequences for bone health after returning to Earth.

- Muscle Atrophy:

 Similarly, the lack of gravity reduces the need for muscles to work against resistance, leading to muscle atrophy.

 This loss of muscle mass and strength can impair an astronaut's ability to perform physical tasks in space and during planetary exploration.

 Muscle atrophy also has implications for overall health, as it can affect metabolism, cardiovascular function, and mobility.

- Cardiovascular Deconditioning:
 In microgravity, the heart does not have to work as hard to pump blood against gravity, which can lead to cardiovascular deconditioning. This condition can result in reduced cardiac output, orthostatic intolerance (difficulty standing up due to low blood pressure), and an increased risk of fainting. Cardiovascular deconditioning can pose serious challenges for astronauts who need to perform physically demanding tasks, such as spacewalks or surface exploration on Mars.

- Fluid Redistribution:

 In microgravity, bodily fluids tend to shift toward the head, causing facial puffiness, nasal congestion, and increased intracranial pressure.

 This fluid shift can also affect vision, leading to conditions such as spaceflight-associated neuro-ocular syndrome (SANS), which can cause changes in eyesight that may persist after returning to Earth.

- Psychological and Neurological Effects:

The absence of gravity may also have psychological and neurological effects, including altered sensory perception, disrupted sleep patterns, and changes in cognitive function.

The impact of microgravity on mental health is an important area of research, as long-duration missions will require astronauts to maintain high levels of cognitive performance and psychological resilience.

Concepts of Artificial Gravity

Artificial gravity is an engineering solution designed to replicate the effects of Earth's gravity within a spacecraft.

There are several concepts for creating artificial gravity, each with its advantages and challenges.

- Rotating Habitats:
 The most widely studied method for creating artificial gravity is through the use of rotating habitats. By spinning a section of the spacecraft or space station, centrifugal force is generated, which can simulate the effects of gravity. The speed of rotation and the radius of the habitat determine the strength of the artificial gravity. This concept has been explored in various designs, including rotating space stations, wheel-shaped spacecraft, and rotating sections within a larger spacecraft.

- Engineering Challenges:

Implementing rotating habitats presents several engineering challenges. One of the main issues is the Coriolis effect, which can cause disorientation and motion sickness as objects and fluids within the rotating environment experience forces that are not present in a non-rotating reference frame. Another challenge is the structural integrity of the rotating habitat, as it must withstand the stresses of rotation while maintaining a safe and stable environment for the crew.

- Human Factors:

 The impact of artificial gravity on human physiology is still not fully understood, particularly in terms of how different levels of gravity (e.g., partial gravity, such as that found on the Moon or Mars) affect health over the long term.

 Research is ongoing to determine the optimal rotation speed and radius for minimizing adverse effects while providing sufficient gravitational force to maintain health.

- Tether Systems:

Another approach to creating artificial gravity involves using tethers to connect two sections of a spacecraft, which are then spun around a common center of mass.

This creates artificial gravity in both sections. Tether systems are relatively simple and lightweight, but they require precise control to maintain stability and prevent the tether from tangling or breaking.

Tether-based artificial gravity systems have been proposed for missions to Mars and other planets, where they could provide a rotating "gravity module" for the crew to use during the journey.

- Hybrid Systems:

 Hybrid systems that combine rotating habitats with traditional microgravity environments could offer a flexible solution.

 For example, a spacecraft might include a rotating section for exercise and sleep, while other activities take place in microgravity.

This approach allows astronauts to experience the benefits of artificial gravity while still taking advantage of the unique conditions of microgravity for scientific research.

Hybrid systems also offer the potential for adjustable gravity levels, allowing astronauts to gradually acclimate to different gravitational forces.

Engineering Challenges of Creating Artificial Gravity

Creating artificial gravity within a spacecraft involves overcoming several technical and engineering challenges.

These challenges include managing the structural integrity of the rotating habitat, ensuring the safety and comfort of the crew, and addressing the potential physiological effects of artificial gravity.

- Structural Integrity:
 The forces generated by a rotating habitat place significant stress on the spacecraft's structure. Engineers must design the habitat to withstand these forces without compromising safety or functionality. This may involve using advanced materials, such as carbon fiber composites, that offer high strength-to-weight ratios. Additionally, the connection points between the rotating and non-rotating sections of the spacecraft must be carefully engineered to prevent mechanical failures.

- Safety and Comfort:
 Ensuring the safety and comfort of the crew is a top priority in the design of artificial gravity systems. This includes managing the Coriolis effect, which can cause disorientation and motion sickness. To minimize these effects, the rotation speed of the habitat must be carefully controlled, and the design should allow for gradual acclimation to the rotating environment. Additionally, the habitat must provide a stable, vibration-free environment for sleeping, eating, and other daily activities.

- Energy Requirements:

 Rotating habitats require energy to initiate and maintain rotation.

 Engineers must balance the energy requirements of artificial gravity systems with the overall energy budget of the spacecraft.

 This may involve optimizing the efficiency of the rotation mechanism and using energy recovery systems to capture and reuse kinetic energy generated by the rotation.

- Physiological Effects:

 The long-term physiological effects of artificial gravity are not yet fully understood.

 Research is needed to determine how different levels of artificial gravity affect bone density, muscle mass, cardiovascular function, and other aspects of human health.

Studies conducted on Earth, such as those using centrifuges to simulate artificial gravity, can provide valuable insights, but space-based research will be necessary to fully understand the impact of artificial gravity in the context of long-duration space travel.

Impact on Human Physiology and Long-Term Health

Artificial gravity has the potential to mitigate many of the adverse health effects associated with long-duration space travel.

However, the impact of artificial gravity on human physiology, particularly over extended periods, is an area of ongoing research.

- Bone Density and Muscle Mass:

 Artificial gravity can help maintain bone density and muscle mass by providing the mechanical loading that stimulates bone formation and muscle activity. Rotating habitats could include exercise facilities where astronauts can engage in weight-bearing activities similar to those on Earth. This could reduce the risk of osteoporosis and muscle atrophy, improving overall health and performance during the mission.

- Cardiovascular Health:

Artificial gravity could also help maintain cardiovascular health by preventing the deconditioning that occurs in microgravity. By simulating the effects of gravity on blood flow and circulation, artificial gravity could reduce the risk of orthostatic intolerance and other cardiovascular issues. This is particularly important for astronauts who will need to perform physically demanding tasks, such as landing on and exploring the surface of another planet.

- Fluid Redistribution:

 Artificial gravity may help mitigate the fluid redistribution that occurs in microgravity, reducing the risk of conditions like spaceflight-associated neuro-ocular syndrome (SANS).

 By encouraging fluids to move toward the lower body, artificial gravity could prevent the buildup of pressure in the head and reduce the risk of vision changes and other related issues.

- Psychological and Cognitive Effects:

The presence of artificial gravity may have positive effects on psychological well-being and cognitive function.

By providing a more Earth-like environment, artificial gravity could help astronauts maintain their circadian rhythms, sleep patterns, and overall mental health.

Additionally, the ability to move and interact with objects in a gravity environment could reduce the cognitive load associated with adapting to microgravity, making it easier for astronauts to perform tasks and make decisions.

8. Autonomous Systems and AI in Space Travel

As space missions become more complex and distant, the need for advanced autonomous systems and artificial intelligence (AI) becomes increasingly important.

These technologies can enhance the safety, efficiency, and success of space missions by enabling spacecraft to operate independently, make decisions in real-time, and adapt to unexpected challenges.

This section explores the role of AI in space travel, including autonomous spacecraft operation, navigation, repair and maintenance systems, health monitoring, and the ethical considerations associated with AI in space exploration.

Role of AI in Spacecraft Operation and Navigation

AI and autonomous systems play a crucial role in the operation and navigation of spacecraft, particularly for missions beyond low Earth orbit, where communication delays with Earth can make real-time control impractical.

- Autonomous Navigation:

 Autonomous navigation systems use AI algorithms to process data from onboard sensors, such as cameras, radar, and lidar, to determine the spacecraft's position, orientation, and trajectory.

 These systems can make real-time adjustments to the spacecraft's course, avoiding obstacles, correcting for deviations, and optimizing the path to the destination.

Autonomous navigation is essential for missions to distant planets, where communication delays make it difficult for ground controllers to provide timely guidance.

- Mission Planning and Execution:

 AI can assist with mission planning by analyzing vast amounts of data to identify the most efficient and effective strategies for achieving mission objectives.

 During the mission, AI systems can execute complex sequences of actions, such as deploying scientific instruments, conducting experiments, and managing resources.

AI's ability to process and analyze data in real-time allows for more flexible and adaptive mission planning, enabling spacecraft to respond to changing conditions and unforeseen challenges.

- Autonomous Landing and Surface Operations:
 For missions involving landing on other planets or moons, AI is critical for ensuring a safe and accurate landing. Autonomous landing systems use AI to analyze terrain data and select the best landing site, taking into account factors such as slope, surface roughness, and the presence of obstacles. Once on the surface, AI can assist with rover navigation, resource extraction, and other surface operations, enabling more efficient and effective exploration.

Autonomous Repair and Maintenance Systems

Spacecraft on long-duration missions must be capable of repairing and maintaining themselves, as the distance from Earth makes it difficult or impossible for human intervention.

Autonomous repair and maintenance systems, powered by AI, are key to ensuring the reliability and longevity of spacecraft.

- Self-diagnosis and Fault Detection:

 AI-powered systems can continuously monitor the health and status of spacecraft components, identifying potential faults or failures before they become critical.
 These systems use data from sensors, such as temperature, vibration, and pressure sensors, to detect anomalies and assess the condition of the spacecraft's systems. Early detection of issues allows for timely intervention, reducing the risk of mission failure.

- Automated Repair Systems: Once a fault is detected, AI-driven repair systems can autonomously diagnose the problem and implement a solution. This may involve using robotic arms, 3D printers, or other tools to repair or replace damaged components. For example, if a micrometeoroid punctures the spacecraft's hull, an automated repair system could deploy a patch or apply a sealant to prevent air loss. AI can also coordinate multiple repair tasks simultaneously, optimizing the use of resources and minimizing downtime.

- Redundancy and Fault Tolerance:
 Autonomous systems can manage redundancy and fault tolerance by automatically switching to backup systems or reconfiguring the spacecraft's operations to bypass faulty components. AI can also analyze the impact of failures on the overall mission and adjust mission plans accordingly. This ability to adapt to unexpected challenges is critical for ensuring the success of long-duration missions, where human intervention may not be possible.

AI in Health Monitoring and Emergency Response

AI has the potential to revolutionize health monitoring and emergency response during space missions, providing real-time support to astronauts and enhancing their safety and well-being.

- Continuous Health Monitoring:

 AI can be used to continuously monitor the health of astronauts, analyzing data from wearable sensors, such as heart rate monitors, blood pressure cuffs, and oxygen sensors. By tracking vital signs and other health indicators, AI can detect early signs of illness, fatigue, or stress, allowing for timely intervention. AI can also analyze trends in health data over time, providing insights into the long-term effects of space travel on human physiology.

- Personalized Health Interventions:

AI-driven health systems can provide personalized recommendations for maintaining physical and mental health during the mission. This may include suggestions for exercise routines, dietary adjustments, or stress management techniques based on the individual's health data. In the event of a health issue, AI can assist with diagnosis and treatment, offering guidance on medication, therapy, or other interventions.

- Emergency Response and Decision Support:

 In the event of an emergency, such as a medical crisis, equipment failure, or environmental hazard, AI can provide real-time decision support to astronauts.

 AI systems can analyze the situation, assess risks, and recommend the best course of action.

For example, if an astronaut experiences a medical emergency, AI can guide their crewmates through the steps of administering first aid or performing a life-saving procedure.

AI can also coordinate the spacecraft's response to emergencies, such as activating alarms, sealing off damaged sections, or adjusting life support systems.

Challenges and Ethical Considerations of AI in Space Travel

The use of AI in space travel raises several challenges and ethical considerations that must be addressed to ensure the safe and responsible deployment of these technologies.

- Reliability and Trust:

One of the primary challenges of using AI in space travel is ensuring the reliability of AI systems. AI must be capable of making accurate and safe decisions, even in complex and unpredictable environments. Building trust in AI systems is critical, as astronauts and mission planners must have confidence in the technology's ability to perform as expected. This requires rigorous testing, validation, and certification of AI systems before they are deployed on missions.

- Transparency and Explainability:

 AI systems must be transparent and explainable, meaning that the reasoning behind their decisions should be understandable to humans.

 This is particularly important in high-stakes situations, where astronauts need to understand the rationale for a particular course of action. Developing AI systems that can provide clear explanations of their decisions and actions is an ongoing area of research.

- Autonomy vs. Human Control:

 The balance between AI autonomy and human control is a key ethical consideration in space travel.
 While AI can enhance the efficiency and safety of missions, there is a risk of over-reliance on autonomous systems, which could lead to reduced human oversight and control. Finding the right balance between AI autonomy and human decision-making is essential for ensuring the success and safety of space missions.

- Ethical Use of AI:
 The ethical use of AI in space travel involves considerations such as privacy, fairness, and accountability. For example, AI systems used for health monitoring must protect the privacy of astronauts' medical data. Additionally, AI systems should be designed to avoid biases and ensure that decisions are fair and just. Establishing ethical guidelines for the use of AI in space exploration is crucial for maintaining the integrity of space missions and protecting the rights and well-being of astronauts.

9. Psychological and Social Aspects of Long-Duration Space Travel

The psychological and social aspects of long-duration space travel are critical to the success of missions that take astronauts far from Earth for extended periods.

The challenges of isolation, confinement, and stress in the harsh environment of space can have significant impacts on mental health and team dynamics.

This section explores the psychological challenges of space travel, the importance of group dynamics, communication with Earth, and the design of spaceships to support mental health and well-being.

Psychological Challenges of Isolation and Confined Spaces

Space travel presents unique psychological challenges, particularly in the context of long-duration missions where astronauts are isolated from Earth and confined to a limited living space.

- Isolation from Earth:

 Astronauts on long-duration missions experience extreme isolation, with limited opportunities for direct contact with family, friends, and the broader human community.

 This isolation can lead to feelings of loneliness, homesickness, and disconnection from the world. The psychological impact of isolation is compounded by the knowledge that help from Earth is distant and delayed, increasing stress and anxiety.

- Confinement:

 The confined living spaces aboard spacecraft limit the physical movement and personal space available to astronauts.

 This confinement can lead to a sense of claustrophobia and a lack of privacy, which can contribute to stress, irritability, and interpersonal conflicts.

 The monotony of the environment, with its lack of variation and natural stimuli, can also lead to mental fatigue and reduced cognitive function.

- Sensory Deprivation:

 The environment of a spacecraft is inherently artificial, with limited sensory input from the natural world. The absence of natural light, fresh air, and the sounds of nature can contribute to sensory deprivation, which can have negative effects on mood, sleep, and overall mental health. Sensory deprivation can also exacerbate feelings of isolation and confinement, making it more difficult for astronauts to cope with the challenges of space travel.

- Stress and Anxiety:

The high-stakes nature of space missions, where every decision and action can have life-or-death consequences, adds to the psychological stress experienced by astronauts. The awareness of being millions of kilometers from Earth, with limited support and resources, can heighten anxiety and fear. The cumulative effects of stress and anxiety can impair cognitive function, decision-making, and emotional regulation, making it more difficult for astronauts to perform their duties effectively.

Group Dynamics and Team Cohesion in Space

The success of long-duration space missions depends not only on individual resilience but also on the ability of the crew to work together effectively as a team.

Group dynamics and team cohesion are critical factors that influence the overall performance and well-being of the crew.

- Team Selection and Training:

The selection of astronauts for long-duration missions involves careful consideration of their psychological compatibility, as well as their technical skills.

Team members must be able to work well together, communicate effectively, and manage conflicts constructively.

Pre-mission training includes team-building exercises and simulations that help astronauts develop trust, cooperation, and problem-solving skills.

Understanding each team member's strengths, weaknesses, and interpersonal dynamics is essential for fostering a cohesive and high-functioning crew.

- Communication and Collaboration:

 Effective communication is vital for maintaining team cohesion and ensuring that the crew can work together efficiently.

 Astronauts must be able to communicate openly and honestly, share information, and provide support to one another.

Communication is particularly important during high-stress situations, where misunderstandings or miscommunications can lead to mistakes or conflicts.

Regular team meetings, check-ins, and debriefings are important for maintaining open lines of communication and addressing any issues that arise.

- Conflict Resolution:

 Conflicts are inevitable in any team, especially in the confined and stressful environment of space.

 However, how conflicts are managed can significantly impact team cohesion and mission success.

Astronauts are trained in conflict resolution techniques, including active listening, negotiation, and compromise.

It is important for the crew to address conflicts promptly and constructively, with a focus on finding mutually acceptable solutions.

Leadership plays a key role in managing conflicts and maintaining a positive team environment.

- Social Support and Morale:

Social support within the crew is essential for maintaining morale and psychological well-being.

Astronauts must be able to rely on one another for emotional support, encouragement, and companionship.

Positive social interactions, shared experiences, and group activities can help build a sense of camaraderie and reduce feelings of isolation.

Maintaining a strong sense of team identity and purpose is also important for boosting morale and helping the crew stay motivated throughout the mission.

Importance of Communication with Earth and Virtual Reality Support

Maintaining communication with Earth is a vital aspect of supporting the psychological well-being of astronauts on long-duration missions.

In addition to traditional communication methods, virtual reality (VR) technology offers new possibilities for enhancing the connection between astronauts and Earth.

- Communication with Family and Friends:

 Regular communication with family and friends on Earth is essential for maintaining emotional well-being and a sense of connection to home.

 Video calls, emails, and messages can help astronauts stay in touch with their loved ones, providing comfort and reducing feelings of isolation.

However, communication delays, particularly for missions to distant planets like Mars, can make real-time conversations challenging.

This delay can exacerbate feelings of isolation, so mission planners are exploring ways to mitigate these effects, such as scheduling regular communication windows and using asynchronous communication tools that allow for the exchange of longer, thoughtful messages.

- Support from Ground Teams:

 Communication with mission control and other support teams on Earth is critical for ensuring the success of the mission and the well-being of the crew.

 Ground teams provide technical support, psychological counseling, and other resources to help astronauts manage the challenges of space travel.

Regular check-ins with psychologists and other mental health professionals can help astronauts cope with stress, anxiety, and other psychological issues that may arise during long-duration missions.

Ground teams also play a crucial role in providing encouragement, problem-solving assistance, and a link to life on Earth, helping astronauts maintain a sense of purpose and motivation.

- Virtual Reality (VR) Support:

 Virtual reality technology offers promising possibilities for enhancing the psychological well-being of astronauts by providing immersive experiences that can simulate Earth-like environments or facilitate social interactions.

 VR can be used to create virtual nature walks, allowing astronauts to experience the sights and sounds of Earth's natural landscapes, which can help reduce stress and combat sensory deprivation.

Additionally, VR can simulate family gatherings or social events, helping astronauts feel connected to their loved ones despite the vast distances.

- Telepresence and Holographic Communication:

 Advanced forms of communication, such as telepresence and holographic technology, are being explored to create more immersive and interactive experiences for astronauts.

 These technologies could enable astronauts to interact with life-like representations of people on Earth, enhancing the sense of presence and emotional connection.

While these technologies are still in development, they hold potential for improving the quality of life for astronauts on long-duration missions.

Designing Spaceships for Mental Health and Well-Being

The design of spacecraft plays a critical role in supporting the mental health and well-being of astronauts.

Thoughtful design can help mitigate the psychological challenges of space travel, promoting a sense of comfort, security, and normalcy in an otherwise extreme environment.

- Personal Space and Privacy:

 Providing astronauts with personal space where they can retreat and have privacy is important for mental health. Individual sleeping quarters or personal modules allow astronauts to have their own space to relax, sleep, and engage in personal activities. These spaces should be designed to be comfortable and customizable, allowing astronauts to personalize their environment with photos, decorations, or other personal items.

- Environmental Control and Lighting:
The environment within the spacecraft, including temperature, lighting, and noise levels, should be carefully controlled to create a comfortable and conducive living space. Lighting systems that simulate natural light cycles can help regulate circadian rhythms, improve sleep quality, and reduce fatigue. Adjustable lighting that allows astronauts to control the brightness and color temperature of their surroundings can also enhance comfort and well-being.

- Biophilic Design Elements: Incorporating elements of nature into the spacecraft design, known as biophilic design, can help reduce stress and improve mental health.

 This could include the use of natural materials, colors, and patterns, as well as the integration of living plants or water features. Even artificial representations of nature, such as images or sounds of forests, oceans, or gardens, can have positive effects on mood and well-being.

- Recreation and Exercise Areas:

 Providing spaces for recreation and exercise is essential for maintaining both physical and mental health. Exercise equipment, such as treadmills, stationary bikes, and resistance machines, should be available to help astronauts stay physically active and counteract the effects of microgravity. Recreational spaces where astronauts can engage in hobbies, play games, or socialize with crewmates can also help alleviate boredom, reduce stress, and promote team cohesion.

- Color and Aesthetics:

The color scheme and aesthetics of the spacecraft interior can influence mood and well-being.

Calming colors, such as blues and greens, can create a soothing atmosphere, while brighter colors can energize and uplift. The overall design should be aesthetically pleasing and functional, creating a positive and stimulating environment for the crew.

10. Ethical and Societal Implications of Interplanetary Travel

As humanity embarks on the journey to explore and potentially settle other planets, a range of ethical and societal implications must be carefully considered. These issues encompass the environmental impact of space travel, the ethical considerations of space colonization, the ownership and use of extraterrestrial resources, and the broader implications for humanity as we become a multi-planetary species.

This section delves into these complex topics, exploring the challenges and responsibilities that come with interplanetary exploration.

Ethical Considerations of Space Colonization

The prospect of colonizing other planets raises profound ethical questions that must be addressed as we plan for the future of space exploration.

- Human Rights and Equality:

 Ensuring that space colonization efforts respect human rights and promote equality is a fundamental ethical consideration.

 Space colonization should not perpetuate existing social inequalities or create new forms of discrimination.

Access to space and its resources should be inclusive and equitable, with opportunities for all people, regardless of nationality, gender, or socioeconomic status.

The governance structures established for space colonies must protect the rights and well-being of all inhabitants, including considerations of freedom, safety, and participation in decision-making.

- Indigenous Rights of Extraterrestrial Life:

 If extraterrestrial life is discovered on other planets, we must consider the ethical implications of human colonization in those environments.

 The potential impact of human activities on indigenous life forms, even if they are microbial, raises important questions about our responsibility to protect and preserve alien ecosystems.

Just as we recognize the rights of indigenous peoples on Earth, we must consider the rights and interests of any extraterrestrial life forms, avoiding actions that could harm or exploit them.

- Sustainability and Environmental Stewardship:

Space colonization efforts must prioritize sustainability and environmental stewardship.

This includes minimizing the ecological impact of human activities on other planets and ensuring that the exploitation of extraterrestrial resources does not lead to irreversible environmental damage.

The principles of sustainability that guide environmental protection on Earth should be applied to space exploration, with a focus on preserving the natural state of other planets for future generations.

Environmental Impact of Space Travel

The environmental impact of space travel extends beyond the confines of Earth, affecting both our home planet and the environments of other celestial bodies.

- Space Debris and Pollution:

The proliferation of space debris in Earth's orbit poses significant risks to both current and future space missions.

Collisions between debris and operational spacecraft can result in further fragmentation, creating a cascade effect that increases the density of debris.

This "Kessler syndrome" could render certain orbits unusable and threaten the safety of space exploration.

Mitigating space debris through improved spacecraft design, active debris removal, and responsible mission planning is essential to ensuring the sustainability of space activities.

- Impact on Planetary Environments:

 Human exploration of other planets, particularly Mars, has the potential to alter their environments in significant ways.

 The introduction of Earth-based microbes, chemicals, and other contaminants could disrupt native ecosystems or interfere with the search for indigenous life.

Planetary protection protocols are designed to minimize these risks, but as exploration and colonization efforts intensify, the potential for environmental impact will increase.

Ethical considerations must guide our approach to preserving the natural state of other planets, balancing the goals of exploration with the need to protect extraterrestrial environments.

- Carbon Footprint of Space Missions:

 Space missions, especially those involving heavy-lift rockets, contribute to the carbon footprint and environmental impact on Earth.

 The production, launch, and operation of spacecraft consume significant amounts of energy and resources, leading to greenhouse gas emissions and other environmental effects.

Developing more sustainable technologies, such as reusable rockets, greener propellants, and more efficient spacecraft, can help reduce the environmental impact of space travel.

Space Law and Ownership of Extraterrestrial Resources

 The legal framework governing space exploration and the ownership of extraterrestrial resources is a complex and evolving area of international law.

- The Outer Space Treaty:

 The Outer Space Treaty, adopted in 1967, is the foundational legal document governing activities in space.

 It establishes that space is the "province of all mankind" and that the exploration and use of outer space shall be conducted for the benefit of all countries.

The treaty prohibits the national appropriation of space by any means, including claims of sovereignty or ownership of celestial bodies.

It also emphasizes the importance of peaceful exploration and the prevention of harmful contamination of space and celestial bodies.

- Resource Utilization and Property Rights:

 As interest in the extraction of extraterrestrial resources, such as water ice on the Moon or minerals on asteroids, grows, questions about property rights and resource ownership are becoming more pressing.

 The Outer Space Treaty does not explicitly address the commercial exploitation of space resources, leading to debates about how these activities should be regulated.

Some countries have enacted national legislation allowing their citizens to claim ownership of resources extracted from celestial bodies, raising concerns about the potential for conflicts and the need for an international legal framework to govern resource utilization.

- Space Governance and International Cooperation:

 The governance of space exploration and resource utilization requires international cooperation and the development of new legal agreements.

 Ensuring that space activities are conducted in a manner that benefits all humanity, rather than a select few, will require the establishment of transparent, inclusive, and equitable governance structures.

These structures should promote the peaceful use of space, prevent the militarization of space, and ensure that the exploration and exploitation of space resources are conducted in a sustainable and responsible manner.

Preparing Humanity for Becoming a Multi-Planetary Species

The idea of humanity becoming a multi-planetary species is both a grand vision and a profound challenge, requiring careful consideration of the social, cultural, and ethical implications.

- Cultural and Societal Adaptation:

 The transition to becoming a multi-planetary species will require significant cultural and societal adaptation.

 This includes developing new cultural norms, values, and traditions that reflect the realities of life on other planets.

Space colonies will need to establish their own social structures, legal systems, and governance models that are adapted to the unique challenges of living in space.

Education and public outreach will play a key role in preparing society for the idea of living on other planets, fostering a sense of shared purpose and responsibility for the future of humanity.

- Ethical Responsibility to Future Generations:

As we embark on the journey to other planets, we must consider our ethical responsibility to future generations.

Decisions made today about space exploration and colonization will have long-lasting impacts on the future of humanity and the environments of other planets.

It is our responsibility to ensure that these activities are conducted in a way that preserves opportunities for future generations, protects the natural state of other planets, and promotes the long-term survival and prosperity of humanity as a whole.

- Interplanetary Governance and Citizenship:

 As humanity establishes settlements on other planets, the concept of citizenship will need to evolve.

 Questions about the rights and responsibilities of space colonists, their relationship with Earth, and the governance structures that will oversee interplanetary colonies are crucial to address.

Will space colonists be considered citizens of Earth or will they hold a distinct planetary citizenship?

How will laws and regulations be enforced across the vast distances between planets?

These questions highlight the need for new frameworks of governance that can adapt to the unique challenges of interplanetary society while ensuring fairness, justice, and equality.

- The Role of Education and Science:

 Education and scientific research will be critical in preparing humanity for life on other planets.

 Understanding the science of space travel, planetary environments, and the technologies needed for survival and thriving in space is essential.

Moreover, fostering a culture of curiosity, exploration, and respect for other worlds will help future generations carry forward the values that will be needed to sustain a multi-planetary civilization.

Investing in space education and research today is an investment in the future of humanity as a multi-planetary species.

11. Conclusion: The Future of Human Space Exploration

The journey to becoming a multi-planetary species is one of the most ambitious and challenging endeavors humanity has ever undertaken.

As we set our sights on the stars, the advancements in space travel technology, spacecraft design, and life support systems will be crucial for ensuring the safety and success of interplanetary missions.

The road ahead is fraught with challenges—ranging from the physical hazards of space travel to the psychological and social aspects of long-duration missions—but it is a journey that holds the promise of extraordinary rewards.

Recap of Technological Advancements Needed

To safely and successfully travel to other planets, we must continue to push the boundaries of technology and innovation.

Key areas of focus include:

- Advanced Propulsion Systems:

 Developing faster and more efficient propulsion technologies, such as nuclear thermal propulsion, ion drives, and eventually fusion or antimatter propulsion, is essential for reducing travel time and making interplanetary missions feasible.

- Radiation Protection:

 Innovative materials and shielding techniques, as well as strategies for minimizing exposure, are critical for protecting astronauts from the harmful effects of space radiation.

- Life Support Systems:

 Reliable, sustainable life support systems that can produce oxygen, recycle water, grow food, and manage waste will be crucial for long-duration missions and space colonization.

- Artificial Gravity:

  Implementing artificial gravity in spacecraft through rotating habitats or other methods will help mitigate the health effects of prolonged microgravity, ensuring the physical well-being of astronauts.

- Autonomous Systems and AI:

 AI-driven autonomous systems will enhance spacecraft operation, navigation, and maintenance, while also providing health monitoring and emergency response capabilities.

- Psychological and Social Support:

 Addressing the psychological challenges of isolation, confinement, and stress, as well as fostering strong group dynamics, will be essential for the mental health and cohesion of the crew.

The Potential Timeline for Achieving Safe Interplanetary Travel

While the exact timeline for achieving safe interplanetary travel is uncertain, significant progress is being made.

NASA, ESA, SpaceX, and other space agencies and companies are actively working on missions to return humans to the Moon and, eventually, to send the first crewed missions to Mars.

The Artemis program aims to establish a sustainable human presence on the Moon by the late 2020s, with Mars missions potentially following in the 2030s.

As technology advances and international cooperation grows, these timelines may accelerate, bringing us closer to the reality of becoming a multi-planetary species.

The Role of International Cooperation and the Private Sector

The future of space exploration will be shaped by both international cooperation and the contributions of the private sector.

Collaborative efforts between space agencies, such as NASA, ESA, Roscosmos, and CNSA, will be essential for pooling resources, sharing expertise, and tackling the challenges of interplanetary travel.

The involvement of private companies like SpaceX, Blue Origin, and others will drive innovation, reduce costs, and expand the possibilities for space exploration and colonization.

The establishment of international legal frameworks and governance structures will be crucial for ensuring that space exploration is conducted peacefully, sustainably, and for the benefit of all humanity.

As we venture into the cosmos, it is important that we do so with a sense of shared responsibility and cooperation, recognizing that space is a common heritage of all humankind.

Inspiring Future Generations of Space Explorers

The dream of exploring and settling other planets has the power to inspire future generations of scientists, engineers, and explorers.

Education and public outreach will play a vital role in fostering a culture of curiosity, innovation, and exploration.

By engaging young people in space science and technology, we can cultivate the skills and passion needed to carry humanity forward into the next great era of exploration.

The vision of becoming a multi-planetary species is not just about survival; it is about expanding the horizons of human knowledge, creativity, and possibility.

As we look to the stars, we must also look within, ensuring that the values of peace, cooperation, and respect for life guide our journey.

The future of human space exploration is a collective endeavor, one that requires the contributions and commitment of people from all walks of life, united by the shared goal of exploring the unknown and building a better future for all.

Please use the next few pages for your notes and debates.

www.ingramcontent.com/pod-product-compliance
Lightning Source LLC
Chambersburg PA
CBHW052139220526
45471CB00004B/1448